FOCUS ON ELEMENTARY

Geology

Teacher's Manual

3rd Edition

Rebecca W. Keller, PhD

Real Science-4-Kids

Illustrations: Janet Moneymaker

Focus On Elementary Geology Teacher's Manual—3rd Edition
ISBN 978-1-941181-41-6

Published by Gravitas Publications Inc.
www.gravitaspublications.com
www.realscience4kids.com

A Note From the Author

This curriculum is designed to provide an introduction to geology for students in the elementary level grades. *Focus On Elementary Geology—3rd Edition* is intended to be used as the first step in developing a framework for the study of real scientific concepts and terminology in geology. This *Teacher's Manual* will help you guide students through the series of experiments in the *Laboratory Notebook*. These experiments will help the students develop the skills needed for the first step in the scientific method — making good observations.

There are several sections in each chapter. The section called *Observe It* helps the students explore how to make good observations. The *Think About It* section provides questions for the students to think about and use to make further observations. In every chapter there is a *What Did You Discover?* section that gives the students an opportunity to summarize the observations they have made. A section called *Why?* provides a short explanation of what students may or may not have observed. And finally, in each chapter there is a section called *Just For Fun* that contains an additional activity.

The experiments take up to 1 hour. The materials needed for each experiment are listed on the next page and also at the beginning of each experiment.

Enjoy!

Rebecca W. Keller, PhD

Materials at a Glance

Experiment 1	Experiment 3	Experiment 4	Experiment 5	Experiment 6
colored pencils	small shovel or garden trowel	baseball or similar hard-centered ball	2 liters (8 cups) or more of dirt for mud pies	a toy, small music box, or toy car that can be taken apart
Experiment 2	small pail or plastic container	balloon	1.75 liters (7 cups) or more of water	a second similar item that can be taken apart
plastic hammer	measuring cup	water	15 milliliters (1 Tbsp.) baking soda	screwdriver
regular metal hammer	dirt that contains rocks (.25 liter [1 cup])	piece of string to tie balloon closed	15 milliliters (1 Tbsp.) vinegar	small hammer
3 pieces of banana	1 tall clear glass container (approx. size: .5 liter [2 cups])	colored pencils	measuring cup	other tools as needed
3 hardboiled eggs in the shell	flour (60 ml [1/4 cup])	**Optional**	measuring spoon	
3 raw potato halves	water	funnel	3 containers (about 1.75 liter [7 cups] size)	
3 rocks of the same type and size (students can collect these)	cake mix and items needed to make the cake		spoon	
safety glasses	nuts, gumdrops, chocolate chips, and/or M&Ms		garden trowel	
Optional			bucket	
8 pieces of paper			paper	
marking pen			marking pen	
			pencil	
			colored pencils	

Experiment 7	Experiment 8	Experiment 9	Experiment 10	Experiment 12
2 clear, tall glasses (drinking or parfait glasses)	outdoor thermometer	3 Styrofoam cups: 355 ml (12 oz.) size	pencil	seeds (student selected)
spoon (1 or more)	helium-filled balloon	about 240 ml (1 cup) each:*	colored pencils	a garden bed or containers and potting soil
3-6 student-chosen food items for parfait model of Earth's layers (such as: graham crackers, peanut brittle, cookies, hot fudge, Jell-O, pudding, ice cream, cream cheese, cherry, nut, jelly bean, etc.)	string	sand	**Experiment 11**	tools for tending plants
		pebbles	2 bar magnets (narrow magnets work best)	herb seeds or small herb plants (student selected)
		small rocks	small, flat-bottomed, clear plastic box (big enough for 2 magnets to fit underneath with some space around them)	
		3 containers for collecting sand, pebbles, and small rocks	corn syrup	
		garden trowel or small shovel	iron filings, about 5 ml (1 teaspoon)**	
		pencil	**Optional**	
student-chosen inedible items that can be used to build a parfait model of Earth's layers (such as: rocks, mud, dirt, clay, dog or cat food, Legos, etc.)		1-2 measuring cups	tape	
		water	2 plastic bags for collecting iron filings	
		enough dirt, pebbles, rocks, water, etc. to make a mud city		
		Optional		
colored pencils		stopwatch or clock with second hand		

* Student-collected or purchased from a place that sells aquarium supplies
** See *Experiment* section for how students can collect iron filings — or iron filings may be purchased at
    www.hometrainingtools.com

Materials: Quantities Needed for All Experiments

Equipment	Materials	Foods
baseball or similar hard-centered ball box, small, flat-bottomed, clear plastic (big enough for 2 magnets to fit underneath with some space around them) bucket containers (about 1.75 liter [7 cups] size), 3 containers for collecting sand, pebbles, and small rocks, 3 garden bed, or containers and potting soil garden trowel glass container, clear, tall (approx. size: .5 liter [2 cups]) glasses (drinking or parfait glasses), clear, tall, 2 hammer, metal hammer, plastic hammer, small magnets, bar, 2 (narrow magnets work best) measuring cup, 1-2 measuring spoons pail, small, or plastic container safety glasses screwdriver shovel, small, or garden trowel spoon thermometer, outdoor tools, misc. as needed tools for tending plants toy, small music box, or toy car that can be taken apart, 2 **Optional** bags, plastic, for collecting iron filings, 2 funnel stopwatch or clock with second hand	balloon balloon, helium-filled cups, Styrofoam, 355 ml (12 oz.) size, 3 dirt for mud pies, 2 liters (8 cups) or more dirt that contains rocks (.25 liter [1 cup]) dirt, pebbles, rocks, water, etc. to make a mud city inedible items, student-chosen, that can be used to build a parfait model of Earth's layers (such as: rocks, mud, dirt, clay, dog or cat food, Legos, etc.) iron filings, about 5 ml (1 teaspoon) [see *Experiment* section for how students can collect iron filings. Or iron filings may be purchased: www.hometrainingtools.com] paper pebbles, about 240 ml (1 cup)* pen, marking pencil pencils, colored rocks, 3 of the same type and size (students can collect these) rocks, small, about 240 ml (1 cup)* sand, about 240 ml (1 cup)* seeds (student selected) seeds, herb, or small herb plants (student selected) string water **Optional** paper, 8 pieces pen, marking tape	baking soda, 15 milliliters (1 Tbsp.) banana, 3 pieces cake mix and items needed to make the cake corn syrup eggs, hardboiled in the shell, 3 flour (60 ml [1/4 cup]) food items, student-chosen, such as: graham crackers, peanut brittle, cookies, hot fudge, Jell-O, pudding, ice cream, cream cheese, cherry, nut, jelly bean, etc. foods, assorted, such as nuts, gumdrops, chocolate chips, and/or M&Ms potato halves, raw, 3 vinegar, 15 milliliters (1 Tbsp.)

* Student-collected or purchased from a place that sells aquarium supplies

Contents

Experiment 1

Geology Every Day

Materials Needed

- colored pencils

Objectives

In this experiment students will explore their surroundings and observe how geology affects their daily lives.

The objectives of this lesson are:

- To encourage students to observe their surroundings.
- To help students explore the different aspects of geology (rock-part, air-part, water-part, and bio-part) and note how these aspects are interconnected.

Experiment

I. Think About It

Read this section of the *Laboratory Notebook* with your students.

Have the students think about where they live. Help them think about their local surroundings, noting the weather, types of wildlife, landforms, and any other features that stand out.

❶-❼ Have them answer the questions in this section. There are no right answers for these questions. Just allow the students to explore their own ideas about the geology of their surroundings.

II. Observe It

Read this section of the *Laboratory Notebook* with your students.

❶-❹ Have your students make a list of all the geological features they see in a day. They can make the list as they travel during the day, or they can take a walk outside and make observations.

They are directed to make lists that include the various types of geological aspects: rock-part, water-part, air-part, and bio-part Have them note features such as parks, trees, lakes, rivers, mountains, the weather, and any other features that stand out.

III. What Did You Discover?

Read the questions with your students.

❶-❹ Have the students answer the questions. These can be answered orally or in writing. Again, there are no right answers and their answers will depend on what they actually observed.

IV. Why?

Read this section of the *Laboratory Notebook* with your students.

Discuss any questions that might come up.

V. Just For Fun

Read this section of the *Laboratory Notebook* with your students.

Help the students think about the various aspects of geology they have explored in this experiment and what those features might be like on the Moon.

Have them draw their ideas and encourage them to use their imagination. There are no right or wrong ideas in this exercise.

Smashing Hammers

Materials Needed

- plastic hammer
- regular metal hammer
- 3 pieces of banana
- 3 hardboiled eggs in the shell
- 3 raw potato halves
- 3 rocks of the same type and size (students can collect these)
- safety glasses

Optional

- 8 pieces of paper
- marking pen

Objectives

In this experiment students will explore how using different tools causes different outcomes.

The objectives of this lesson are to have students:

- Explore how different objects have different properties.
- Observe how the properties of different materials require the use of different tools.

Experiment

Before beginning the experiment have the students collect three rocks of the same type and similar size. The experimental results will vary depending on the hardness of the rocks and whether or not they are layered.

I. Think About It

Read this section of the *Laboratory Notebook* with your students.

Have the students think about the differences between a plastic hammer and a metal hammer. Also have them think about the differences between a piece of banana, a hardboiled egg in its shell, a raw potato, and a rock. They may answer the questions orally or in writing.

Guide open inquiry with questions such as:

- *Is a metal hammer heavier or lighter than a plastic hammer? Why?*

- *Do you think a hardboiled egg is harder or softer than a banana? Why or why not?*

- *Is a rock harder or softer than a hardboiled egg? Why?*

- *Do you think a plastic hammer can crush a rock? Why or why not?*

- *Do you think a metal hammer can crush a hardboiled egg? Why or why not?*

- *Do you think a metal hammer would work better for smashing a potato than a plastic hammer would? Why or why not?*

II. Observe It

Read this section of the *Laboratory Notebook* with your students.

In this experiment students will observe how using different tools can result in different outcomes. They will first smash a piece of banana with a plastic hammer and then smash another piece of banana with a metal hammer. This process will be repeated with hardboiled eggs in the shell, potato halves, and rocks. Students can write or draw their observations in the boxes provided, or they can relate their observations orally.

Have the students wear safety glasses to protect their eyes from possible debris.

Optional: Students can place each object on a piece of paper before smashing it and write or draw which hammer they'll be using.

Guide open inquiry with questions such as:

- *What happens to the piece of banana when you smash it with the plastic hammer? What does it look like after it's smashed? Why?*

- *What happens to the piece of banana when you smash it with the metal hammer? What does it look like? Does the hammer you use make a difference? Why?*

- *Does it make a difference whether you smash a hardboiled egg with the plastic hammer or the metal hammer? If so, what is different?*

- *Can you smash the potato with the plastic hammer? With the metal hammer? Is there a difference? Why or why not?*

- *What happens when you smash the rock with the plastic hammer? With the metal hammer? Does one of the hammers work better? Why or why not?*

III. What Did You Discover?

Read this section of the *Laboratory Notebook* with your students.

Have the students answer the questions in this section orally or in writing. Have them refer to their notes or drawings in the *Observe It* section. There are no right or wrong answers to these questions.

IV. Why?

Read this section of the *Laboratory Notebook* with your students.

Discuss any questions that might come up.

V. Just For Fun

Read this section of the *Laboratory Notebook* with your students.

Have the students use a magnifying glass or hand lens to look at each of the objects that is unsmashed, that was smashed with the plastic hammer, and that was smashed with the metal hammer. Students may write, draw, or state their observations orally.

Guide open inquiry with questions such as:

- *Did [the object] smash the way you thought it would? Why or why not?*

- *Does [the object] look different when you look at it through the magnifying glass (or hand lens) than it does when you look at it using only your eyes? Why or why not?*

- *Does the unsmashed [object] look different from the way the smashed one looks? How?*

- *Can you see a difference between the way [the object] looks when it was smashed with a plastic hammer and when it was smashed with a metal hammer? Why or why not?*

Experiment 3

Mud Pies

Materials Needed

- small shovel or garden trowel
- small pail or plastic container
- measuring cup
- dirt that contains rocks
  (.25 liter [1 cup])
- 1 tall clear glass container
  (approx. size: .5 liter [2 cups])
- flour (60 ml [1/4 cup])
- water
- cake mix and items needed to
  make the cake
- nuts, gumdrops, chocolate
  chips, and/or M&Ms

Objectives

In this experiment the students will explore how sedimentary rocks are made. The students will experiment with the sedimentation process by using a mixture of rocks, dirt, water, and flour.

The objectives of this lesson are for the students to:

- Observe the differences between rocks and dirt.
- Perform an experiment that illustrates the sedimentation process.

Experiment

I. Observe It

Read this section of the *Laboratory Notebook* with your students.

❶ Help your students collect a sample of dirt that contains rocks from their backyard, park, or some other place that will allow a sample to be taken.

❷ Use questions such as the following to help your students make good observations as they look through the dirt sample.

- *What does the dirt and rock mixture look like?*
- *Can you describe materials in this mixture that are different from each other?*
- *What do you think the different materials are?*

❸ Have your students use their hands to separate the rocks from the particles of soil. The rocks can be large or small.

❹ Guide your students' inquiry with the following questions:

- *What do the rocks look like?*
- *Can you describe their shape, color, or feel (texture)?*
- *What does the dirt look like?*
- *Can you describe the shape, color, or feel (texture) of the dirt?*
- *How can you tell the difference between a rock and dirt?*

❺ Help your students make a slurry of rocks, dirt, and water. There should be sufficient space in the clear glass container that a layer of water 5-8 centimeters (2-3 inches) thick can form on top of the rock and soil mixture.

❻ Guide your students' inquiry with the following questions:

> • *What happens to the rocks when you swirl the mixture?*
>
> • *What happens to the dirt when you swirl the mixture?*
>
> • *What happens to the water when you swirl the mixture?*

❼ Have the students allow the mixture to settle and then help them record their observations.

II. Think About It

Read this section of the *Laboratory Notebook* with your students.

❶-❷ Have your students answer the questions. Have them base their answers on their actual observations and not on what they think should have happened.

Use the following questions to help your students think about why layers form.

> • *Do you think rocks are heavier or lighter than dirt? Why or why not?*
>
> • *Does the water make the rocks and dirt settle to the bottom more quickly or more slowly? Why or why not?*
>
> • *What do you think would happen if you added oil instead of water to your mixture? Would the rocks settle more quickly or more slowly? Why or why not?*

Help your students see that layers are formed as heavier objects settle first and lighter objects settle last. The water slows down the settling process because water is thicker than air. Oil would slow it down even further since oil is thicker than water.

❸ Have the students add 60 ml (1/4 cup) of flour to the slurry and stir it. The flour will mix with the lighter dirt particles and may form a muddy-white layer when the mixture has settled.

Have the students record their observations.

III. What Did You Discover?

Read the questions with your students.

❶-❹ Discuss the questions in this section with the students. Their answers may vary since the answers are based on what they actually observed. There are no right or wrong answers to these questions.

IV. Why?

Read this section of the *Laboratory Notebook* with your students.

Explain to the students that this experiment shows how sedimentary rocks form. The word sediment comes from the Latin word *sedimentum* which means "to settle" and refers to solid matter that settles out of a fluid such as water or air. Sedimentary rocks begin to form when sediments settle and form layers as some sediments settle faster than others. When the layers of sediment are put under pressure, sedimentary rocks are formed.

What your students observed was exactly this process of sedimentation. The heavier rocks settled first, followed by the lighter rocks and dirt, leaving a layered sediment. Most sedimentary rocks are formed in this fashion.

V. Just For Fun

Read this section of the *Laboratory Notebook* with your students.

A great way to illustrate the process of sedimentation is to make a cake with layers of heavier nuts, gumdrops, chocolate chips, and/or M&Ms. Most recipes ask that you put flour on the additions to keep them from falling to the bottom. However, for this recipe do not use the flour so that students can observe how the heavier items settle to the bottom of the pan during the baking process.

Experiment 4

The Shape of Earth

Materials Needed

- baseball or similar hard-centered ball
- balloon
- water
- piece of string to tie balloon closed
- colored pencils

Optional

- funnel

Objectives

In this experiment students will explore how Earth gets its shape. Earth is almost spherical but bulges slightly at the equator. The central bulge is created by centrifugal force, an outward force caused by the rotation of Earth on its axis.

The objectives of this lesson are for students to:

- Observe the difference between a hard sphere (baseball) and a liquid-centered "sphere" (water balloon).
- Explore how the creation of centrifugal force by the Earth's rotation creates a central bulge.

Experiment

I. Think About It

Read this section of the *Laboratory Notebook* with your students.

Have the students think about the shape of Earth and why it is shaped like a slightly smashed ball. There is no right answer to this question. Allow the students to explore their own ideas about the cause of Earth's shape.

II. Observe It

Read this section of the *Laboratory Notebook* with your students.

❶ Have your students observe what happens when they spin a hard ball, such as a baseball, on the floor. Have them note whether the ball changes shape.

❷ Have them record their observations.

❸ Help the students fill a balloon with water and tie it closed. Putting the opening of the balloon over a garden hose works well, or you can use a funnel. Have the students observe what happens when they spin a fluid-centered ball, such as the water balloon, on the floor. Have them note whether or not the water balloon changes shape as it spins.

❹ Have them record their observations.

III. What Did You Discover?

Read this section of the *Laboratory Notebook* with your students.

❶-❹ Have the students answer the questions. These can be answered orally or in writing. Again, there are no right answers, and their answers will depend on what they actually observed.

IV. Why?

Read this section of the *Laboratory Notebook* with your students.

Discuss any questions that might come up.

V. Just For Fun

In this section students are introduced to thought experiments which are performed by using the imagination to think through the possibilities of a theory rather than by doing a physical experiment. Einstein was a famous scientist who used thought experiments to lead to new discoveries in physics.

Students are asked to imagine what it would be like if Earth were cube-shaped instead of spherical. Help spark your students' imagination with questions such as the following:

- *What do you think would happen if you were riding in a car and you came to an edge of the Earth?*

- *Would it make a difference if the edges of the Earth were sharp or if they were rounded?*

- *What would it be like if your house was near a corner of the Earth?*

- *What would the oceans be like?*

- *What would Earth look like if you were high up in an airplane?*

There are no right answers. Encourage students to use their imagination freely.

Have them record their ideas.

Mud Volcanoes

Materials Needed

- 2 liters (8 cups) or more of dirt suitable for making mud pies
- 1.75 liters (7 cups) or more of water
- 15 milliliters (1 tablespoon) baking soda
- 15 milliliters (1 tablespoon) vinegar
- measuring cup
- measuring spoons
- 3 containers for mixing mud (about 1.75 liter [7 cups] size)
- spoon or other implement for mixing mud
- garden trowel
- bucket
- paper
- marking pen
- pencil
- colored pencils

Objectives

In this experiment students will make different mixtures of dirt and water to explore how the viscosity of lava determines the type of volcano formed.

The objectives of this lesson are for students to:

- Observe the differences between thick and thin mud mixtures.
- Observe how different mud mixtures will form different types of mounds and then compare these results to the way different types of volcanoes are formed.

Experiment

I. Think About It

Have the students think about the questions in this section. They can write their answers in the space provided or they can give their answers orally. There are no right or wrong answers to these questions.

Use questions such as the following to guide open inquiry.

- *When you pour syrup on your pancakes, what happens? Why do you think this happens?*

- *How fast does the syrup move? Why?*

- *How would the syrup act if you poured so much of it on your pancakes that it ran off the edge of the plate? Why?*

- *If you poured water on your pancakes, do you think it would be different from pouring syrup on them? Why or why not?*

- *How fast would the water move? Why?*

- *Is there a difference in how thick and thin liquids move? Why or why not?*

- *What do you think you would need to make a good mud pie?*

II. Observe It

Have the students dig up dirt that is suitable for making mud pies, or provide it for them. They will need at least 1.5 liters (6 cups) of dirt for this part of the experiment and .5 liter (2 cups) or more of dirt for the *Just For Fun* section.

Part I

Have the students follow the directions in the *Laboratory Workbook* to make 3 separate mud mixtures. Have them label each container of mud as **A**, **B**, or **C**.
A marking pen can be used or the students can come up with their own ideas for labeling.

If needed, have the students adjust the dirt:water ratio so that mixture **A** is like a thick paste, mixture **B** is somewhat thinner, and mixture **C** is more liquid.

[Parts II–IV: A table follows each set of questions in **Parts II-IV**. Have the students answer the questions by filling in the appropriate boxes in the table, or they may answer orally. Answers will vary and there are no right or wrong answers.]

Part II

Have the students observe the thickness of each mixture. Use questions such as the following to guide inquiry.

- *What happens when you add more water to the dirt? Why?*
- *Which mud mixture is easiest to mix? Hardest? Why?*
- *Do you think if you make a mud pie from each of the mixtures they will all look alike? Why or why not?*

Part III

Have the students select three separate areas in which to pour the three different mud mixtures. Each area should be labeled **A**, **B**, or **C** with a piece of paper or in some other way.

In general, students should observe that mixture **A** is thicker and more difficult to pour than liquid **C** and that mixture **A** will form a higher mound while mixture **C** will spread out the most.

Use questions such as the following to guide inquiry.

- *Do the different mixtures pour differently? In what ways?*
- *Do all the mixtures spread out in the same way after they are poured? Why or why not?*
- *Which mixture covers the largest area? Why?*

Part IV

In this part of the experiment students will add more layers of mud to each mound, letting the mud pie dry out each time before adding another layer of mud. They are to pour the same mixture in the same spot each time. They should observe that liquid **A** forms a more compact mound than liquid **B** or **C** and that liquid **A** takes fewer layers to build up the mound.

Use questions such as the following to guide inquiry.

- *Do the mixtures all make mounds of the same height with the same number of layers? Why or why not?*

- *Which mound is the widest? The highest, the middle height, or the lowest? Why?*

- *What do you observe about the relationship of height to width in the different mounds? Why do you think they are different?*

III. What Did You Discover?

Have the students answer the questions in this section. In general they should observe that the thicker mud of mixture **A** would make a cone volcano shape and the thinner mud of mixture **C** would make a shield volcano shape.

IV. Why?

This experiment helps students explore how different types of lava will form different types of volcanoes. Thick lava that does not flow very far away from the vent where it comes out of the Earth forms cone volcanoes, and thinner lava that can flow very long distances forms shield volcanoes.

V. Just For Fun

Have the students build a mock cone volcano using a thick mud mixture. After they form a mound with the mud, they can use a pencil to poke a hole down the center of the mound while the mud is still wet.

When the mud has dried, help the students measure and pour 15 milliliters (1 tablespoon) of baking soda and then 15 milliliters (1 tablespoon) of vinegar into the center of their volcano and observe what happens. Direct them to avoid looking directly into the volcano opening when they pour the vinegar in. They can record their observations by drawing them in the box provided.

Experiment 6

All the Parts

Materials Needed

- a toy, small music box, or toy car that can be taken apart
- a second similar item that can be taken apart
- screwdriver
- small hammer
- other tools as needed

Note: The objects used in this experiment may not work again.

Objectives

In this experiment, students will explore how taking apart an object helps them learn in detail how the object works.

The objectives of this lesson are for students to:

- Use suitable tools to study an object.
- Observe how tools help scientists make better observations.

Experiment

Before starting the experiment

This experiment requires that the students disassemble an object to learn more about the individual parts. For this experiment choose a toy or small mechanical object that is OK to disassemble, understanding that the object may not work again when reassembled. Make sure the item is one that has a number of pieces that can be taken apart.

Inspect the object to be disassembled to see what tools will need to be provided to the students.

I. Think About It

Read this section of the *Laboratory Notebook* with the students.

Help the students think about the parts of a bicycle, a car, and an airplane. Discuss which parts are similar (e.g., wheels) and which parts are different (e.g., wings, motor, gears).

Explore open inquiry with questions such as the following:

- *What parts can you think of that are on a bicycle?*
- *What parts can you think of that are on a car?*
- *What parts can you think of that are on an airplane?*
- *Which parts do you think are similar and which do you think are different?*
- *Are there any parts that you think you would you find on all three objects? What are they?*
- *Are there any parts that you think you would find only on the bicycle? Only on the car? Only on the airplane?*

II. Observe It

Read this section of the *Laboratory Notebook* with your students.

❶ Have the students carefully observe the object. Guide them to note what the object does and how it functions.

In the box provided, have the students fill in the name of the object and write and/or draw their observations.

❷ Have the students carefully take apart the object, and guide them in making observations during disassembly. Have the students do as much of the disassembly themselves as possible. Once the object is taken apart, have the students count the number of parts and write it on the line provided.

❸ Students are to examine each part. Boxes are provided for them to draw the parts and to record the size, weight, and apparent function of each. Relative weights can be used.

❹ Have the students reassemble the object. The object may or may not work when they are finished. Have them record their observations.

III. What Did You Discover?

Read this section of the *Laboratory Notebook* with your students.

Have the students answer the questions. There are no right answers and their answers will depend on what they actually observed

IV. Why?

Read this section of the *Laboratory Notebook* with your students.

Discuss any questions that might come up.

V. Just For Fun

Have the students take apart another object. Guide them in making good observations, and have them record their observations in the boxes provided.

Experiment 7

Edible Earth Parfait

Materials Needed

- 2 clear, tall glasses (drinking or parfait glasses)
- spoon (1 or more)
- 3-6 student-chosen food items that can be used to build a parfait model of Earth's layers (such as: graham crackers, peanut brittle, cookies, hot fudge, Jell-O, pudding, ice cream, cream cheese, cherry, nut, jelly bean, etc.)
- student-chosen inedible items that can be used to build a parfait model of Earth's layers (such as: rocks, mud, dirt, clay, dog or cat food, Legos, etc.)
- colored pencils

Objectives

In this experiment, students will explore how models help scientists make educated guesses about how things work.

The objectives of this lesson are for students to:

- Use suitable tools to study an object.
- Observe how tools help scientists make better observations.

Experiment

I. Think About It

Read this section of the *Laboratory Notebook* with your students.

Help your students think about different foods they could use for the crust, lithosphere, asthenosphere, mesosphere, outer core, and inner core. Students may combine layers with similar qualities. Solid, thin food items like crackers, peanut brittle, or cookies might be good items for the crust. Softer food items like hot fudge, Jell-O, pudding, or ice cream might be good items for the inner layers. A cherry, nut, jelly bean, or similar item could be used for the inner core. Help your students think about the consistency of different foods and whether or not they might make a good representation of a particular layer of an edible Earth.

Explore open inquiry with the following:

- *What are some solid food items you like?*

- *What are some soft food items you like?*

- *What are some combinations of food items you like?*

- *How well do you think different food items will fit together and taste?*

- *Which layers of Earth do you want to include in your parfait? All of them? If not all, which ones? Why?*

- *Which layers would you represent with solid (hard) food? Which would you make with softer foods? Why?*

II. Observe It

Read this section of the *Laboratory Notebook* with your students.

❶ Help the students plan a layered Edible Earth Parfait. This experiment requires that the students use a variety of food items. Guide students to pick foods that work for you and for them. Have them list the foods and which layer each food represents.

❷ Have students think about whether they were able to find foods to represent the properties of the different the layers of Earth.

❸-❹ Have the students assemble their Edible Earth Parfait and make careful observations about it. Help them observe whether the layers are interacting.

❺ The students can now eat their model of Earth's layers.

❻ Have the students look up the definition of parfait in a dictionary or online. Ask them if they think their edible Earth model fits the definition of a parfait. Why or why not?

III. What Did You Discover?

Read this section of the *Laboratory Notebook* with your students.

Have the students answer the questions. There are no right answers, and their answers will depend on what they actually observed.

IV. Why?

Read this section of the *Laboratory Notebook* with your students.

Discuss any questions that might come up.

V. Just For Fun

Have the students review the *Think About It* section and choose inedible items to create another model of Earth's layers. A box is provided for them to draw and label their model. Have them observe any similarities or differences between the edible and inedible models. Are the layers interacting in this model?

Experiment 8

What's the Weather?

Materials Needed

- colored pencils
- outdoor thermometer
- helium-filled balloon
- string

This experiment is done over the course of a week.

Objectives

In this weeklong experiment students will explore the atmosphere by making daily observations of the weather and its effects.

The objectives of this lesson are for students to:

- Observe daily changes in the weather.
- Observe how changes in the weather can affect plants, animals, and the land.

Experiment

I. Think About It

Read this section of the *Laboratory Notebook* with your students.

Have the students think about what the weather is like where they live and how it changes from day to day and from season to season. Guide open inquiry using questions such as:

- *What do clouds look like? Do they always look the same? Why or why not?*

- *Do you think changes can happen to the ground when it rains hard? Why or wy not?*

- *What do you think happens to plants when there is a breeze? When the wind blows really hard? Why?*

- *Do you think wind can make changes to the land? Why or why not?*

- *Do you think animals act differently when it is raining than when it is sunny? When it is very windy? When it is very hot or very cold? When it is day or night? Why or why not?*

- *Do you think you act differently in different kinds of weather? During the daytime or at night? In summer and winter? Why or why not?*

II. Observe It

Read this section of the *Laboratory Notebook* with your students.

Have the students observe the weather at about the same time every day for a week. Have them record the temperature, and write, draw, or relate orally their observations about the weather. Guide open inquiry with questions such as:

- *What do you see when you look at the sky? Is it the same color every day? Why or why not?*

- *What do you notice about clouds? What do they look like? Do they look the same every day? Why or why not?*

- *Is the temperature the same every day? Is it the same at night? Why or why not?*

- *What differences do you observe in animal behavior as the weather changes? When it's sunny? When it rains? When it's windy?*

- *Do changes in the weather affect the plants? When it's sunny? When it rains? When it's windy? Why or why not?*

- *Do you observe any changes to the ground when it rains hard or is very windy? Why or why not?*

III. What Did You Discover?

Read this section of the *Laboratory Notebook* with your students.

Have the students review their notes from the *Observe It* section and answer the questions based on their observations. There are no right answers.

IV. Why?

Read this section of the *Laboratory Notebook* with your students.

Discuss any questions that might come up.

V. Just For Fun

Read this section of the *Laboratory Notebook* with your students.

In this experiment students will make a simple tool to measure the wind.

Help the students find an object outdoors that they can tie the balloon to. The balloon should have enough space around it that it won't hit anything if the wind blows hard. Have the students observe the balloon several times during the day.

Guide open inquiry with questions such as:

- *Do you think you will be able to test the wind by using a balloon? Why or why not?*

- *What things do you think the balloon might be able to measure?*

- *By looking at the balloon, can you tell if the wind is blowing? Why or why not?*

- *Can you tell how hard the wind is blowing? Why or why not?*

- *Can you tell the direction of the wind? Why or why not?*

- *Do you think a balloon is a good way to measure the wind? Why or why not?*

Experiment 9

How Fast Is Water?

Materials Needed

- 3 Styrofoam cups: 355 ml
  (12 oz.) size
- about 240 ml (1 cup) each:*
  sand
  pebbles
  small rocks
- 3 containers for collecting
  sand, pebbles, and small rocks
- garden trowel or small shovel
- pencil
- 1-2 measuring cups
- water

  * student-collected or
    purchased from a place that
    sells aquarium supplies

Optional

- stopwatch or clock with second
  hand

Just For Fun

- enough dirt, pebbles, rocks,
  water, etc. to make a mud city

Objectives

This experiment introduces students to the concepts of porosity and groundwater. Students will explore how the porosity of a material determines how quickly water can flow through it and how porosity affects the absorption of rainwater into the soil.

The objectives of this lesson are for students to:

- Observe how different materials have different porosities.
- Observe how the porosity of a material affects how water flows through it.

Experiment

I. Think About It

Read this section of the *Laboratory Notebook* with your students.

Have the students think about what happens to rain after it falls to the ground. Guide open inquiry with questions such as:

- *Where do you think the water in lakes and rivers comes from? Why?*

- *What do you think happens to rain after it falls on the land? Where does it go? Why?*

- *Do you think there is a difference between rain falling on the ground and snow falling? Why or why not? What is different?*

- *Do you think you would see a difference between rain that falls on rocks and rain that falls in a garden? If so, what differences would you see?*

- *Why do you think rain sometimes makes mud puddles and sometimes does not?*

- *What do you think keeps the water in a lake?*

- *Do you think plants and animals could live on land where there is no rain or snow? Why or why not?*

II. Observe It

Read this section of the *Laboratory Notebook* with your students.

If possible, have the students collect the sand, pebbles, and small rocks to be used in this experiment.

❶ Have the students use a pencil to poke a hole in the bottom of each Styrofoam cup. The holes should be about the same size.

❷-❹ Have the students measure about 240 milliliters (1 cup) each of the sand, pebbles, and small rocks and pour each into its own cup.

❺-❼ The students are to pour 120 milliliters (4 ounces) of water into each of the three cups and observe how fast the water flows through each material.

The objective of this part of the experiment is for students to compare the relative speed at which water travels through the three different materials. Students can measure the length of time by using a stopwatch or a clock with a second hand, or simply by observing how quickly or slowly water runs through each material compared to the others.

Guide the students to observe how much of the water comes out the bottom of each cup and whether the amount varies by material. They can check it visually or catch the water in a measuring cup.

Have them record their observations.

III. What Did You Discover?

Read this section of the *Laboratory Notebook* with your students.

Have the students review their notes from the *Observe It* section and answer the questions based on their observations. There are no right or wrong answers.

IV. Why?

Read this section of the *Laboratory Notebook* with your students.

Discuss any questions that might come up.

V. Just For Fun

By building a mud city with a landscape that has rivers and a lake, students can begin to explore how surface water and groundwater operate. Students can use various natural materials to build their city—dirt, sand, pebbles, rocks, water, etc. They can try different things like slowing the speed of the river water by using pebbles or rocks or speeding it up by making a smooth riverbed and a higher elevation for the source of the river. A dam can be built to contain the lake water. Allow the students to play with the materials and see what they can discover on their own.

Help the students think about how they want to build their city by asking questions such as:

- *Where do you think the city part should go? Do you think the city would have a wall around it?*

- *Where will the rivers be?*

- *How will you make the water flow in the rivers?*

- *Where will the river water end up?*

- *Can you make the river water move faster or slower?*

- *Where will you put the lake?*

- *How will you keep the water in the lake?*

- *If the water doesn't stay in the lake, where do you think it will go?*

Students can make paper boats to float in their rivers and lake. If they'd like, they can draw their mud city.

Experiment 10

What Do You See?

Materials Needed

- pencil
- colored pencils

Objectives

In this experiment students will explore their environment by walking around their yard and neighborhood, observing plants, insects, animals, and people and any activities they are involved in.

The objectives of this lesson are for students to:

- Observe living things and the environment they live in.
- Observe some of the factors that create that environment.

Experiment

I. Think About It

Read this section of the *Laboratory Notebook* with your students.

By having students think about the living things they will observe when they go out into their environment, they will begin to understand that the living things and the resources within an environment are interconnected and that environments differ from region to region.

Guide open inquiry with questions such as the following.

- *If you walk around outside, what animals, birds, and bugs do you think you will see?*

- *What will the animals, birds, and bugs be eating? Why? What will they be doing? Why?*

- *Where do you think different animals, birds, and bugs sleep? Why?*

- *Do you think all living things sleep? Why or why not?*

- *Do you think you will see the same plants everywhere you look? Why or why not?*

- *Can you think of any animals that need to live near or in water?*

- *Do you think you will see any people? What do you think they'll be doing? Why?*

- *Do you think you'll see places where people can live and places where they can't live? Why or why not?*

II. Observe It

Read this section of the *Laboratory Notebook* with your students.

Have the students walk around their yard and neighborhood, observing what's around them. Encourage them to take their time and to look at things close up as well as from a distance. In addition to observing living things, they can also be guided to notice the weather, the presence or absence of water, the soil, and other factors that go into creating a particular environment.

Guide open inquiry with questions such as:

- *What are the animals, birds, and bugs doing?*

- *What are they eating?*

- *Are they moving around? How?*

- *Are any of them sleeping? Where?*

- *Is the weather making any difference in what the animals and birds are doing? Why or why not?*

- *Are the plants different in different areas? Why or why not?*

- *Are there different animals in different areas? Why or why not?*

- If there is a watery place that can be observed—*Can you tell if there are plants or animals living in the water? What are they doing? What do they eat? Could they live without the water? Why?*

- *Do you see any people? What are they doing? Why?*

Have the students bring their *Laboratory Notebook*, a pencil, and colored pencils with them on their walk so they can make drawings and notes as they make their observations.

The blank line in each heading on the observation pages can be filled in with the location of the observations or some other identification.

III. What Did You Discover?

Read this section of the *Laboratory Notebook* with your students.

Have the students refer to the drawings and notes they made while on their walk. Answers are based on their observations, and there are no right or wrong answers to these questions.

IV. Why?

Read this section of the *Laboratory Notebook* with your students.

Discuss any questions that might come up.

V. Just For Fun

Since the Kepler spacecraft was launched in 2009, scientists have discovered that many stars have planets orbiting them. Those planets that exist outside our solar system are called exoplanets.

The Circumstellar Habitable Zone is the area of a solar system where an Earth-like planet would be at the right temperature to have liquid water and therefore might have the right conditions for life as we know it to exist. Kepler-62e is a recently discovered Earth-like exoplanet in a Circumstellar Habitable Zone and may have conditions suitable for life, although that is not yet known as of this writing.

In this experiment students are to use their imagination to think about what life might be like on another planet. There are no right or wrong answers. Encourage students to go where their imagination takes them, even if their ideas are improbable, wild, and fanciful.

For more about planets and exoplanets, see *Focus on Elementary Astronomy—3rd Edition.*

Experiment 11

Moving Iron

Materials Needed

- 2 bar magnets (narrow magnets work best)
- small, flat-bottomed, clear plastic box
  (big enough for 2 magnets to fit underneath with some space around them)
- corn syrup
- iron filings, about 5 ml
  (1 teaspoon)
  (see Experiment section for how students can collect iron filings—
  or iron filings may be purchased:
  www.hometrainingtools.com)

Optional

- tape
- 2 plastic bags for collecting iron filings

Objectives

In this experiment students will "see" the magnetic field around a magnet.

The objectives of this lesson are for students to:

- Visualize the magnetic field surrounding a magnet.
- Observe how a magnetic field can influence objects that are attracted to magnetic force.

Experiment

Students can gather iron filings themselves by putting a magnet in a plastic bag and dragging the bag through some dirt. Iron that is in the dirt will collect on the outside of the bag. Place the bag containing the magnet inside another plastic bag and then remove the magnet from the inner bag. The iron filings will fall into the outer bag. Repeat several times until about 5 ml (1 teaspoon) of iron filings has been collected.

Alternatively, iron filings can be purchased.

I. Think About It

Read this section of the *Laboratory Notebook* with your students.

Have the students think about magnets and what they have learned about them. Guide open inquiry with questions such as:

- *Do you think everything made of metal can be a magnet? Why or why not?*

- *If you look at a magnet, do you think you will see the magnetic field that surrounds it? Why or why not?*

- *If you were an astronaut in space, do you think you could see Earth's magnetic field? Why or why not?*

- *Do you think magnets can be useful? Why or why not? What could they be used for?*

- *What do you think would happen if you pushed two magnets together matching their north poles? Matching the north pole of one with the south pole of the other? Why?*

II. Observe It

Read this section of the *Laboratory Notebook* with your students.

❶ Have the students pour corn syrup into the box. There needs to be enough syrup that the iron filings can move around freely. A layer of about 6 millimeters (1/4") works well.

❷ Have the students place the box on top of the magnet so that the magnet is centered. The box needs to be flat (not tipped).

   Optional: The magnet can be taped to the bottom of the box.

❸ Have the students pour the iron filings on the syrup, being careful not to breathe them in.

❹ After about 30 minutes the students will be able to observe how the magnetic field of the magnet has affected the iron filings. The corn syrup is not affected by the magnetic field, but the iron filings will have aligned to the magnetic forces.

❺ Have students record what they observe.

III. What Did You Discover?

Read this section of the *Laboratory Notebook* with your students.

Have the students answer the questions based on their observations. Guide open inquiry with questions such as:

- *Do you think the magnetic field of a magnet will always stay the same? Why or why not?*

- *Why do think you had to wait 30 minutes before recording the pattern made by the iron filings?*

- *Do you think a different magnet would make the iron filings move into a similar pattern? A different pattern? Why or why not?*

- *Do you think Earth's magnetic field might have a pattern that is similar to that of the bar magnet? Why or why not?*

IV. Why?

Read this section of the *Laboratory Notebook* with your students.

Discuss any questions that might come up.

V. Just For Fun

❶   Have the students move the box so the magnet is repositioned. Depending on the interest of the student, the magnet can be repositioned more than once. Students will need to wait about 30 minutes each time before making their observations.

Have them record the resulting pattern and observe whether repositioning the magnet changed the pattern of the magnetic field.

❷-❸ In this part of the experiment students will place two magnets under the box to see if adding a magnet changes the magnetic field. Depending on their interest, they can reposition the magnets more than once.

Have them record the resulting pattern(s) and observe whether using two magnets changed the pattern of the magnetic field and also if the pattern was different when the two magnets were moved to new positions.

Experiment 12

What Do You Need?

Materials Needed

- seeds (student selected)
- a garden bed or containers and potting soil
- tools for tending plants
- herb seeds or small herb plants (student selected)

This experiment is done over the course of several weeks.

Objectives

The experiment will take place over the course of several weeks. Students will grow a plant and observe how all the parts of the Earth are interdependent and necessary for plant life.

The objectives of this lesson are for students to:

- Observe factors that are necessary for a plant to grow and be healthy.
- Observe how all the parts of the Earth work together to provide everything needed for a plant to grow.

Experiment

I. Think About It

Read this section of the *Laboratory Notebook* with your students.

Have the students think about the role each part of the Earth plays in the growth of a plant. Guide open inquiry with questions such as:

- *Do you think the atmosphere contains anything that a plant needs in order to grow? Why or why not?*

- *Do you think plants can live without water? Why or why not?*

- *How do you think plants get water from the hydrosphere?*

- *Do you think plants need anything from the biosphere? Why or why not?*

- *Where do you think soil comes from? Why?*

- *Do you think plants could live if there were no magnetosphere? Why or why not?*

II. Observe It

Read this section of the *Laboratory Notebook* with your students.

❶ Help the students choose a plant to grow from seed. For the purposes of this experiment, a plant likely to be eaten by bugs would be a good choice, and one that produces vegetables that can be eaten could be fun to grow. If there's enough space, students might like to grow more than one type of plant.

❷ Have the students plant several seeds in case some of them don't germinate. They may need to thin out some of the plants if too many seeds sprout. Seeds may be planted in a garden bed or in a container in a sunny spot. It's best to have the container outside where the plant can interact with the weather, bugs, etc. If it isn't possible to have the plant outdoors, have the students think about what factors might be different for a plant that is indoors and what advantages and disadvantages an indoor plant might have.

❸ Help the students remember to water the seeds after they're planted and check the soil moisture daily until the seeds sprout. Then have them check the soil on a regular basis. Have them notice in what ways the weather affects the soil.

❹ Students are to write and/or draw any observations they make after they plant the seeds and as they water them. Boxes are provided in the workbook.

❺-❼ Have the students make frequent observations of their plant. Use questions such as the following to guide them in making observations about how different parts, or spheres, of the Earth are affecting their plant.

- **Biosphere:**
  Do you think any other living things are affecting your plant? What are they and what effect are they having?
  (For example, are bugs or rabbits eating the plant, worms enriching the soil, a cat rolling against the plant?)
  What things are you doing that affect the plant?

- **Atmosphere:**
  Do you think wind is affecting your plant? How?
  Is it too sunny, too cloudy, too rainy—or not enough of any of these? How can you tell?

- **Hydrosphere:**
  Is the plant getting enough water from rain (also involves the atmosphere)? Where do you think the water from the hose or spigot comes from? How and why?
  Can you tell when the plant needs more water? Why or why not?

- **Geosphere:**
  Do you think the soil affects how often the plant needs water (also involves the hydrosphere)?
  What does it look like the soil is made of?

- **Magnetosphere:**
  Do you think the magnetosphere is letting the right amount of the Sun's energy reach the plant? How can you tell?

Have the students check the growth and health of their plant frequently and write and/or draw what they observe.

III. What Did You Discover?

Read this section of the *Laboratory Notebook* with your students.

Have the students review their observations and use them to answer the questions. Discuss any questions that may arise. There are no right or wrong answers.

IV. Why?

Read this section of the *Laboratory Notebook* with your students.

Discuss any questions that might come up.

V. Just For Fun

Students are to grow an herb garden of several plants. It can be in a garden bed or in containers outdoors or indoors.

Help the students select herbs to grow. They can research herbs in the library or online or go to a nursery to look at plants. Starting with either seeds or small plants is suitable for this experiment. Students might choose herbs they know they like or ones that look pretty or have intriguing names. Or they may come up with their own selection criteria.

Discuss possible locations for the herb garden and have the students plant the seeds or repot the plants, if needed,

As the herbs are growing, have the students make observations about how the plants grow and what is needed to keep them healthy. The students can decide when to harvest leaves and observe how this affects the plants. If herbs suitable for making tea are grown, students can try cutting off some branches of the herb and hanging them upside down from a string tied around the stems. Once the leaves are dry, they can be put in hot water to make tea. Students can also use this method to dry herbs for later use in foods, storing the dried leaves in airtight containers.

Students who like doing research might enjoy discovering which plants have flowers that are edible and then grow and eat some of these.

More REAL SCIENCE-4-KIDS Books
by Rebecca W. Keller, PhD

Building Blocks Series yearlong study program — each Student Textbook has accompanying
Laboratory Notebook, Teacher's Manual, Lesson Plan, Study Notebook, Quizzes, and Graphics Package

Exploring Science Book K (Activity Book)
Exploring Science Book 1
Exploring Science Book 2
Exploring Science Book 3
Exploring Science Book 4
Exploring Science Book 5
Exploring Science Book 6
Exploring Science Book 7
Exploring Science Book 8

Focus On Series unit study program — each title has a Student Textbook with accompanying
Laboratory Notebook, Teacher's Manual, Lesson Plan, Study Notebook, Quizzes, and Graphics Package

Focus On Elementary Chemistry
Focus On Elementary Biology
Focus On Elementary Physics
Focus On Elementary Geology
Focus On Elementary Astronomy

Focus On Middle School Chemistry
Focus On Middle School Biology
Focus On Middle School Physics
Focus On Middle School Geology
Focus On Middle School Astronomy

Focus On High School Chemistry

Super Simple Science Experiments

21 Super Simple Chemistry Experiments
21 Super Simple Biology Experiments
21 Super Simple Physics Experiments
21 Super Simple Geology Experiments
21 Super Simple Astronomy Experiments
101 Super Simple Science Experiments

Note: A few titles may still be in production.

Gravitas Publications Inc.
www.gravitaspublications.com
www.realscience4kids.com